First published December 1951

First facsimile edition 2008

Copyright © Old Pond Publishing Ltd 2008

All rights reserved. No parts of this publication may be reproduced, stored in a retrieval system, or transmitted, in any form or by any means electronic, mechanical, photocopying, recording or otherwise, without prior permission of Old Pond Publishing.

ISBN 978-1-905523-94-8

A catalogue record for this book is available from the British Library

Published by

Old Pond Publishing Ltd
Dencora Business Centre
36 White House Road
Ipswich IP1 5LT
United Kingdom

www.oldpond.com

The publishers wish to acknowledge with thanks
the help given by Michael Thorne of the Coldridge Ferguson Collection, Devon.

Originally designed and produced by Cogent Advertising Services Ltd, Coventry
This edition designed by Liz Whatling and printed by 1010 Printing Ltd, China

Peter and Pauline at Hollyhock Farm

by
R. A. E. LINNEY

Published by
HARRY FERGUSON LIMITED
COVENTRY, ENGLAND
1951

PETER *and* PAULINE ON THE FARM

PETER and PAULINE were twins. They were exactly nine years old and lived with their mother and father in a nice house in a busy street of a huge city. Daddy was an architect, and was away at the Office all day, so naturally the twins didn't see much of him, but they knew he worked very hard, poor dear, so when they were at home from school they behaved themselves very well and helped their Mummy as much as they could.

One morning at breakfast time, just as they'd finished their porridge, and Mummy had gone into the kitchen to fetch the boiled eggs, Daddy looked up with a smile. He'd been sorting through his mail and it was clear that he'd found a very special letter for he held a large green sheet of notepaper in his hand.

'How would you like to go and live on a nice big farm for three or four weeks?' he asked.

The twins looked at one another in delighted astonishment. A farm — in the country — there'd be pigs — and sheep — and cows — and dogs — and dear old hardworking horses. There might even be all sorts of wonderful machinery too,

such as tractors and lorries and things. The twins were a little vague, really, as to all that happened on a farm, but they were certainly willing to learn.

Hearing all the excitement, Mummy came in from the kitchen with the boiled eggs on a tray.

'What's going on?' she said, looking at the two flushed and excited faces before her.

'It's old Arthur,' said Daddy, smiling at her astonishment. 'He's written and asked if we could spare the twins for a week or so. He knows they're on their holidays and thinks it would do them good to have a few days on his farm.'

'Jolly good idea,' answered Mummy, 'but I'm not really sure that we *can* spare them, are you, Daddy?'

For a moment the twins looked at one another in dismay then seeing the twinkle in their mother's eye their happy laughter rang out afresh.

Meanwhile Mummy and Daddy fell to discussing details and since Uncle Arthur said the twins could come as soon as they liked, Daddy promised to take the day off and drive them to the station that very afternoon.

My goodness! What a hustle and a bustle, a flurry and a scurry there was in Peter and Pauline's house that morning!

There were clothes to pack, ration books to be fetched and all the hundred and one things that were likely to be needed when two healthy youngsters were going for a holiday on a Sussex farm.

At last they were ready, and after a hurried lunch at which both of them were almost too excited to eat anything the time arrived for the luggage to be carefully stowed in the car and the trip through the busy streets to the station began.

Then the last fond farewells were spoken, there was a big kiss each from Mummy and a nice new pound note each from Daddy, and the great adventure had started!

Soon the train came rolling through the broad green downs of Sussex and with their faces glued to the window the twins discussed every passing farmhouse with a newly acquired interest which was almost professional in its enthusiasm.

At long last the train drew in to Fernley Halt, a tiny wayside station fragrant with the loveliness of a warm July evening, and there — on the platform — stood Uncle Arthur and Auntie Mabel with the dog Bess barking excitedly at their heels.

With gleeful shouts the twins ran to them and such a fuss was

made, the like of which Fernley Halt had never seen before.

It wasn't until they were driving along the dusty Sussex roads in their Uncle's car — with their luggage packed in the back seat with them — that there was really time to talk.

'I'm so glad that your Daddy decided to let you come, my dears,' said Auntie Mabel, 'we were beginning to get worried until he telephoned at tea time. Apparently they'd both been

so busy getting you on the train that neither of them had remembered to let us know!'

Dusk was beginning to fall as the happy party drove through the big white gates of Hollyhock, Uncle Arthur's hundred acre mixed farm nestling cosily in the heart of the downs.

In the distance could be heard the faint lowing of cattle, the eager barking of dogs and those happy unknown noises which are for ever a part of the farmer's night.

The car pulled up at the farmhouse door and the little guests got out, sleepy but oh! so excited with the prospect of their new home. Bess frisked round them with glee. She knew that a car trip after the day's work was done was something *most* unusual and whatever it was Bess was fully determined not to miss anything.

While Aunty Mabel was unlocking the door the twins gazed around them. In a corner of the yard — as though waiting for further orders — stood three handsome grey tractors. Peter clapped his hands with joy — real tractors! What more could a boy want?

At this moment Uncle Arthur returned from putting away the car. He patted Peter gaily on the head.

'I see you're admiring our Ferguson tractors,' he said. 'Well,

tomorrow you shall both learn to drive one — that is, if you'd like to.'

In one voice the twins assured him that nothing would please them better, and the cheerful party passed into the spacious kitchen of Hollyhock farmhouse.

It was an old house built in the late seventeenth century, and the big oak beams were dull with the smoke of a thousand kitchen fires. No fire glowed now in the massive brick fireplace for this was July and the scent of apple and roses hovered outside the open windows.

Auntie Mabel bustled about preparing supper for her husband and her young guests. First, a heavy damask table cloth was spread over the freshly scrubbed table, the crockery was laid — some of it had been in the family for generations — a huge cheese and a dish of golden butter were fetched, delicious home-made bread was brought out, a cut glass dish of pickled walnuts was added, together with a bowl of fresh trifle. Milk from Uncle Arthur's own cows supplied the finishing touches of an excellent supper to which the twins, let it be said, did ample justice.

That night as the silver stars twinkled over the gentle hills of the Sussex downs and a pale rind of a moon stippled the mellow tiles of the ancient house of Hollyhock, two little curly heads snuggled down on their pillows to dream of the excitements to come on the morrow.

IN THE WISHING WELL FIELD

THE next morning dawned clear and bright, and the sun, smiling warmly through the windows, wakened the twins with a start and jumping out of bed they rushed to the open window to survey the scene.

How different it was to the busy street which was their home in the city! Here were no double-decker buses, screaming lorries, hooting motor-cars, and careless pedestrians. Here were the gentle sights and sounds of an English farm.

From somewhere out of sight came the merry chugging of a tractor, the clatter of milkchurns, the rattle of pans and the hoarse voices of the workers as they went about their daily tasks. Far out, on the side of the hill, long fields of wheat were ripening in the sun and in the air the skylarks rose like rockets to sing their ecstatic melodies.

In a few seconds the twins had thrown on their clothes and were racing each other downstairs to see who could be first at the pump.

Hands washed, faces washed and hair combed, they sat at the breakfast table while Auntie Mabel told them of the plans for the day.

'Uncle Arthur is down the Wishing Well field with Freddy Burbage and two of the Ferguson tractors, mowing. Your Uncle asked if you would like to go down there and join him.

He says you will find it very interesting and I'll pack your lunch and you can take Bess with you if you like.'

So, soon after breakfast the twins set off, a luncheon basket under one arm and Bess scampering delightedly at their heels.

A few minutes walk along a carefully explained path brought them to the edge of Wishing Well field. This was a huge field so called, their Aunt had explained, because at one time a deep well had existed somewhere in the field though where it was now no-one knew — not even old Gaffer Brownlow, who had worked at Hollyhock for nearly eighty years. Auntie Mabel said that it was most likely filled in and forgotten though if they did find it they might as well have a wish!

At the end of the field, chugging cheerfully in the sun were the two Ferguson tractors driven by Uncle Arthur and Mr. Burbage. Some kind of cutting instruments, the twins could see, were attached to the rear of these tractors and the long grass was being mown at a surprising rate.

They walked slowly down to the end of the field and putting their basket carefully in the shade of the hedge bottom

they turned to wave to Uncle Arthur. He stopped his tractor and came over to them.

'Well, so you're up at last,' he smiled with a friendly pat on the head for the eager Bess and a big kiss for the excited twins. 'And I see you've brought the lunch too, well done!'

He turned and waved to Freddy Burbage to come across.

'How would you like to do some mowing for us?' he said, looking slowly at the open-mouthed twins.

They gazed at one another in astonishment. This was too good to be true. But surely Uncle Arthur couldn't mean it. In any case, how on earth could two nine year old youngsters operate such complicated machinery?

'Its quite easy, you know,' explained Uncle, smiling at the twins' astonishment. 'You see, with one of these Ferguson tractors, operating an attached implement such as a mower is very simple indeed. Mr. Ferguson, who invented this type of tractor, says that even a child could drive one — and I'm sure he's right!'

Taking Peter by the hand, his uncle led him towards the stationary tractor and mower, and in the meantime Freddy Burbage — with a bright 'Top of the morning to you'— had joined the party.

'We'll start with the tractor,' said Uncle, lifting Peter into the sprung-steel seat. 'First of all, we switch on and a red light shows, telling us that the ignition system is ready to fire the cylinders, and that all we have to do is to start the engine.'

Peter nodded. He'd seen Daddy start their own car so many times that he knew all this. But he didn't say anything because he thought it would be very bad manners to interrupt. Besides — all this was jolly interesting anyway!

'Now,' went on Uncle Arthur. 'Here is a very special safety device on this type of tractor — we start the engine by

using the gear lever in the centre. This means that the engine must be in what is known as "neutral," that is, the drive is disconnected from the wheels — before we can start. To relieve any strain on our engine we press down the foot-clutch (there it is, Peter, on the left), push the gear lever up and the engine starts. Try it!'

Peter rather nervously pressed down the clutch, moved the gear lever, and sure enough the engine purred into life while the mower blades rattled merrily behind in unison. This was grand! Pauline was looking on excitedly. She wanted to try too, but thought it wiser to let her brother fully understand things first — after all he'd be able to explain everything to her later.

'Since we're mowing today,' said Uncle, 'we shall use second gear — there, you see that little number two marked on the top of the gear box. All we do to start off is press out our clutch once more, put the gear lever in the gear position selected, open the throttle a little — that's the throttle lever just under the steering wheel — let our clutch come slowly back and off we go.'

'But what about the mowing?' asked Peter, 'how can I control that? I can't keep stopping and getting out of my seat, can I?'

'I'm coming to that,' said Uncle, laughing, 'and I want you to listen carefully. This tractor and the mower behind forms part of what is known all over the world as the FERGUSON SYSTEM. I haven't got time to explain it to you now, but I'll tell you all about it after tea tonight — I know you'll be

interested. Anyway, by this System the implement — which, as you can see, happens to be the mower this morning — can be perfectly controlled by means of the hydraulic lift. This is the hydraulic lift lever on the right, and with it we can raise or lower any implement attached to the rear-end of our tractor.'

'To start the mower, we engage it in gear, as if we want our knife-blades to be run from the tractor engine. Therefore, we move the Power Take-off lever on the left and there we are.'

The engine was started again, the mower blades rattled and they were ready for the off.

'Now,' said Uncle, 'away you go. Take it nice and steady and let me see you mow this little strip down here!'

Gingerly Peter eased in the clutch and off the whole outfit moved forward, followed by the barking, happy Bess with Pauline clapping her hands with glee.

A few yards down the field Peter stopped, pressing out the clutch, applying the handy foot-brake and pulling the gear lever back into the neutral position. He looked round, bursting with pride and self-satisfaction.

'Well done,' shouted Uncle, 'and now I'm going off with Freddy into the next field to do some mowing there. See you at lunch-time. But there's one other thing — the one of you which isn't driving must be very careful indeed not to get into the way of the cutter-bar. It can be very dangerous. So be careful!'

And with a wave of his hand off he went with Freddy Burbage, and a few seconds later the two of them drove gaily out of the Wishing Well field on the other tractor.

The Gear Lever

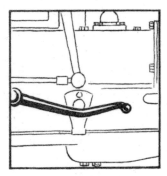

The Clutch Pedal

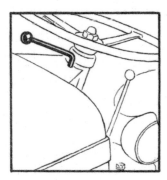

The Throttle Control

The Hydraulic Lift Lever

All that morning under the blazing sun the twins toiled away at their mowing. In fact, by lunch-time they were so proficient in the art of starting, stopping, slowing and turning that the two of them were finally convinced that there was no other tractor in the world like the Ferguson.

A fine lunch they had too, sitting with Uncle and Freddy in the shade of a fine big tree, with Bess wagging her tail and going from first one of them and the other for the bits.

For Auntie had packed them all a most delicious lunch. There were sandwiches and sausage rolls, cheese-cakes and buttered buns, tomatoes and a beautiful jam pasty. And when every crumb had been eaten and washed down with draughts of home-made lemonade from the big stone jar, they agreed that a farmer's life was indeed a happy one.

The afternoon passed just as pleasantly and much useful work was done by Peter and Pauline and the Ferguson tractor deep in the Wishing Well field.

At last six o'clock rolled round, the voices of the tractors stilled and the party went wearily homewards for a well-earned tea. Uncle, of course, had to work again after tea, but Auntie insisted that the twins couldn't possibly mow any more. She was sure they'd had quite enough!

It was after the tea things had been cleared away and the men settled down with their pipes — Gaffer Brownlow gripping his old clay firmly but affectionately in his toothless gums — that Uncle Arthur started to tell them something about the Ferguson System.

'Mr Ferguson,' he explained, 'had realised for a very long time indeed that unless much more food was produced many of us throughout the world would not have enough to eat, and would most likely starve to death. You see,' said he, 'while the population of the world increases from year to year, food production doesn't. In other words, in order that we shouldn't starve, we must produce more food. Another important point is this,' he went on, puffing at his briar, 'if more food was produced the prices of basic essentials would be reduced, and, therefore, in the long run, the cost of living would come down.'

He tapped out his pipe on his boot.

'There is only one way to produce more food, Mr. Ferguson argued. And that is to mechanise the whole system of agriculture — not only in England but throughout the world!'

He looked round the table at the faces around him, all listening with great interest to the views of a real farming man talking about the job he loved best of all — farming.

'Now, a lot of farmers used to employ horses to pull their various implements — their ploughs, their harrows, their mowers and such like. But a horse eats as much as he earns and besides that, he is slow and unreliable. Yet a lot of small farmers couldn't afford to pay the large prices required to buy expensive tractors and trailed equipment. It was too dear. So Mr. Ferguson conceived the idea of building an all-purpose tractor to which the implement could be attached or — as we call it — integrated.'

'Now you must realise that such a novel system of power farming didn't just happen. It needed careful thought and very patient experiment before it could be brought to a pitch where it could be successfully offered to farmers all over the world.'

'You see, Mr. Ferguson first thought of his scheme back in the dark days of the first World War, when he was undertaking a survey of Irish farming for the Irish Department of Agriculture. He was apalled at the slow laborious and near-obsolete methods used by many farmers to cultivate their land. And he

realised, too, that not only were farmers in Ireland using these painfully slow means to raise their crops, but farmers throughout the world were in the same position, struggling along in the manner of their forefathers, to grow food for a world population which was increasing yearly by leaps and bounds. Something had to be done — at once. Or world starvation threatened each and every one of us.'

'He'd always been interested in mechanical things, had Mr. Ferguson — in fact, he designed, built and flew his own aeroplane in

Ireland right back in 1909. But he combined with his enthusiasm for engineering a passionate love for the land and all that for which it stood.'

'The earth is fruitful enough to feed us all, he thought, and some means must be found to enable all farmers — whatever their acreage and whatever the extent of their holdings — to produce their crops more easily and above all — faster!'

'But you can't hurry Nature,' Peter chipped in, with a puzzled face.

'Of course not,' explained his uncle patiently, when the chuckles of laughter had subsided, 'but at least it is possible to get better farming results from the seasons if the problem is tackled the right way. So much, you see, in farming, depends on the weather. Therefore it follows that the more work that can be done when the weather is fine the more crops a farmer can expect from his acres. But, as Mr. Ferguson was wise enough to see, more manpower was *not* the answer. The answer

lay in the world wide application of mechanised farming.'

'So Mr. Ferguson and his associates toiled and experimented through the years to devise a machine which while it could, in effect, be used for almost every operation on the farm, yet had to be marketed at a price which the small farmer could afford. In other words, the tractor pioneered by Mr. Ferguson was not just an ordinary tractor at all — it was the mechanical heart of a great new System of mechanised farming, now known all over the world as the Ferguson System.'

'Basically, the idea is beautifully simple. Using the tractor as the prime power unit, a whole range of farming implements could be mounted — one at a time, of course — on the rear of the tractor and actually be fully controlled and operated by the man on the tractor driving seat. Such a system of implement-integration has an enormous number of advantages, which I'll try to explain to you as simply as I can.'

'For instance, when an implement is towed — or trailed, as we call it — by a tractor, certain natural forces are allowed to struggle against one another.'

'For example, the soil resists the implement strongly — often so strongly that the implement tries to pull the back of the tractor down and thereby the front wheels are forced into the air. You follow me, don't you?'

The twins nodded their heads in agreement.

'In such a system it is often necessary to load the tractor with extra weight, etc., so that it can keep its front end down. But all these extra weights mean extra work for the tractor

to move itself and not much power is left over sometimes for pulling the implements. Besides, heavy tractors pack the soil down hard so that crops won't grow well in it.'

'Now in the Ferguson System, where the implement is actually attached to the tractor by a method of three point linkage, the combined tractor-and-implement use nature's forces to *help* them, not to hinder them!'

'You see, as the tractor travels along, the soil again resists the passage of the implement which — were it being towed instead of carried — would pivot forward. Yet by the action of the top link on the hydraulic mechanism — which I told you about in the field this morning — this particular force is made to put more weight on the back wheels, giving more pulling power and on to the front wheels, keeping the nose of the tractor down. It also keeps the implement at an even depth.'

'So,' interrupted Peter, 'the implement keeps the same depth all the time, until you lift it up or push it deeper by the hydraulic lever?'

'Exactly,' said his Uncle, 'and I ought to say a word or so about the three-point linkage by which all Ferguson implements are attached to the tractor. You see, as the implement tries to pivot forward, the top link of the three-point attachment naturally tries to move forward and downwards — passing a really powerful thrust through the tractor to the wheels. The front wheels are therefore held down very firmly indeed and there is no chance of a front end lift or rear wheel spin, which would make tractor driving very awkward indeed.'

He finished speaking and looked at the twins, as Auntie Mabel butted in.

'Now then!' she said, 'time you were off back to the fields and letting these poor children have a rest.'

'Oh, please Auntie', begged Pauline, 'don't let them go yet. We were so interested.'

'Tell them how they make tractors at Coventry,' chuckled Gaffer Brownlow, who, at ninety-odd thought he could still

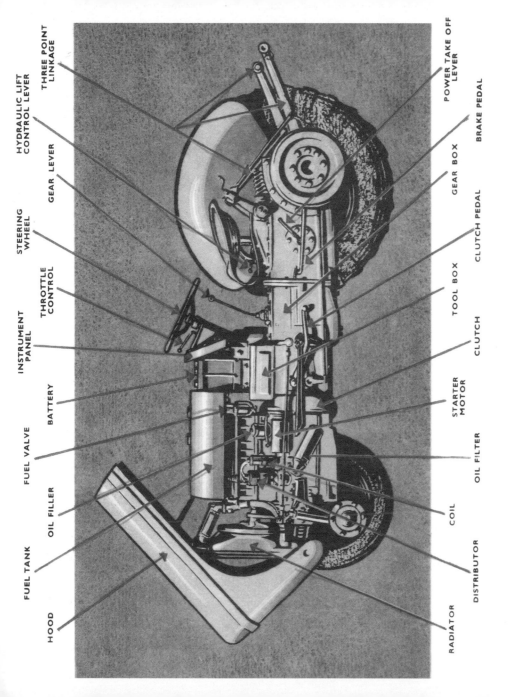

afford to chuckle at these 'darned new fangled devices'!

'I've been to Coventry,' put in Freddy Burbage, his red face and bald head gleaming in the evening sun, 'so I can tell 'em about that.'

'At Coventry — a mighty busy town in what they call the Midlands — you'll find the offices and headquarters of Harry Ferguson Ltd., the company which market the tractors. They don't make them themselves, nor the implements, they merely design them and distribute them all over the world through Dealers and that sort of thing.'

'You see, the Ferguson System has been found to be so good, and the demand for tractors and implements so great that Ferguson's just couldn't make them all themselves, even if they wanted to.'

'The tractors are made in one of the huge factories of the Standard Motor Company Limited, of Coventry, and this factory is situated in Banner Lane.

Resistance of the soil to the implement tends to lift the front wheels.

'It is a very large factory and has been re-tooled and adapted specially for the production of Ferguson tractors. The most modern machinery has been installed, working conditions are good and the whole plant is worth many millions of pounds. And that's where each single Ferguson tractor — sold in this half of the world — is made.'

'What about the other half?' asked the practical Peter.

'There is a factory in Detroit which caters for the Western Hemisphere,' went on Freddy, who seemed to know all about it, 'and the American farmers are very pleased indeed with the Ferguson tractor. As a matter of fact, Mr. Ferguson was selling his tractor in the United States a long time ago, but that's another story and I haven't time to tell you about it now.'

'Also at Coventry — and you can go there yourself if you'd like to ask them — is the most wonderful exhibition I've ever seen. It's all about farming through the ages and all over the world, and it shows how the Ferguson System can save agri-

The Ferguson System keeps the tractor down and the implement at even depth.

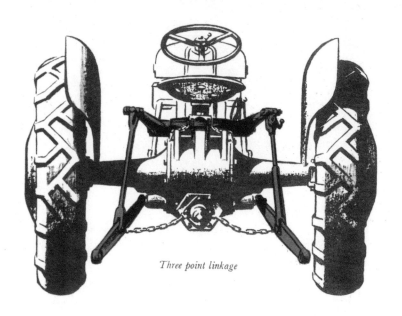
Three point linkage

culture by applying modern mechanical principles. There are some wonderful models, pictures, tableaux and all the latest types of Ferguson Tractors and equipment are on show.'

'And whereabouts in Coventry is this marvellous Exhibition?' asked the wondering twins.

'It's been built right in the offices of the Ferguson Company on Fletchamstead Highway at Coventry,' answered Freddy, 'and thousands of people have visited it already. It was opened in March, 1949, by Lord Boyd-Orr, who is a world-renowned food expert and a very famous man.'

'Where are the implements made then?' asked Peter.

'There are thirty different kinds of Ferguson farming implements, made by many first-class engineering firms up and down the country,' said Freddy thoughtfully, 'as well as many accessories. In fact, there is a specialised implement for almost every type of job that we could encounter on the farm — isn't there Arthur?'

Uncle Arthur nodded.

'And can all of these implements be attached and work on the System — I mean, can they be attached and operated entirely from the driving seat of the tractor?' asked Pauline.

'Nearly all of them,' replied Freddy. 'There is a whole range of these implements you know — ploughs, harrows, cultivators, mowers, steerage-hoes, that are used primarily in the field. Then there are such things as wood-saw, the transport box, the trailers and the earth-scoop whose use is perhaps more general rather than specialised.'

'By using Ferguson tractors combined with this equipment, anyone, even on a tiny five or six acre farm, can grow his food

Coventry — the famous city of the Three Spires.

COVENTRY

From the Headquarters of Harry Ferguson Ltd. at Coventry, the tractors and implements are sent to every part of the Eastern Hemisphere.

STONELEIGH

At the famous old Stoneleigh Abbey, students from all over the world attend the Ferguson School.

DETROIT

In America there is another Ferguson Company, Harry Ferguson Incorporated, which produces tractors and implements for the Western Hemisphere. Between them, Harry Ferguson Ltd. and Harry Ferguson Incorporated, send agriculture equipment to more than 80 countries.

Opposite. This is the Banner Lane factory at Coventry of The Standard Motor Co. Ltd., who make the famous

much more cheaply and much more quickly. Greater yield per acre is the ultimate aim of the Ferguson Plan — and you don't have to be on a 100 acre farm like Hollyhock in order to achieve results.'

The sun was gradually sinking beyond the windows and the men were getting up to go back to their work. Pipes were being knocked out and pocketed and brows — still damp with the sweat of hard toil — were being mopped for the last time before going out of doors. How peaceful it was, thought the twins, as their eyes drank in the homely scene which they were to remember for the rest of their lives.

'One last question,' gasped Peter, and the menfolk turned round from the open door. 'With all this new machinery and the different method of working it, how does a farmer learn about it in the first place?'

Uncle Arthur smiled.

'Well, its not very difficult to operate the Ferguson System, as you yourself have shown. But the ins and outs of the System have to be thoroughly mastered, its true, and the Ferguson Company have thought of a very good idea indeed to teach the farmers all about it.'

'They realised it would be impossible to come and instruct each farmer and his workers individually. Therefore, they have set up a big Training School at a beautiful place called Stoneleigh Abbey in Warwickshire. It's quite near to Coventry and a very fine organisation has been set up there to teach Ferguson Dealers and their Staff — and through them, the farmer and his staff, all about the Ferguson System.

'Students have attended this School from countries all over the world, to take back to their own countrymen the story of the Ferguson System. For here at the school, a complete ten-day Field Course is held — as well as a practical service course dealing with the servicing and operation of the tractors and implements.'

'Can anyone go to the Ferguson School?' asked Peter hopefully, wide-eyed with anticipation.

Freddy chuckled, and even Gaffer Brownlow couldn't resist a wheezy laugh.

'I don't think so,' said Uncle kindly, 'you see, really the School is intended only for Ferguson representatives, Ferguson Dealers and their staffs. It's only fair after all, because it would be impossible to take everybody and by doing it in the way they do the Ferguson Company think that farmers everywhere will be able to learn at first hand from these people who have attended the School.'

And with that the menfolk clattered out into the heavy perfumed stillness of the evening, each bent on completing as much of his task as he could before the dusk fell and a velvet-like darkness covered the downs.

That night as they lay luxuriously tired in their bed, the heavy-eyed twins discussed in a low voice the excitements of their first day at Hollyhock.

They remembered with pleasure the brown, faithful eyes of dear old dog Bess, the kindly toothless smile of ninety-four year old Gaffer Brownlow, the jokes of Freddy Burbage and the masculine kindness of Uncle Arthur.

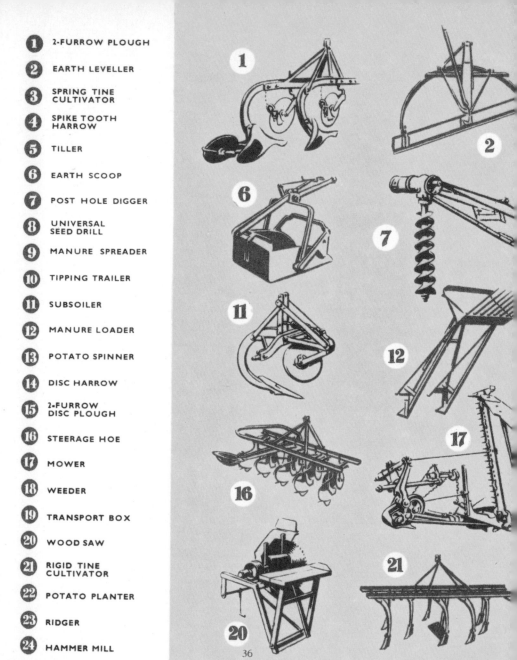

1. 2-FURROW PLOUGH
2. EARTH LEVELLER
3. SPRING TINE CULTIVATOR
4. SPIKE TOOTH HARROW
5. TILLER
6. EARTH SCOOP
7. POST HOLE DIGGER
8. UNIVERSAL SEED DRILL
9. MANURE SPREADER
10. TIPPING TRAILER
11. SUBSOILER
12. MANURE LOADER
13. POTATO SPINNER
14. DISC HARROW
15. 2-FURROW DISC PLOUGH
16. STEERAGE HOE
17. MOWER
18. WEEDER
19. TRANSPORT BOX
20. WOOD SAW
21. RIGID TINE CULTIVATOR
22. POTATO PLANTER
23. RIDGER
24. HAMMER MILL

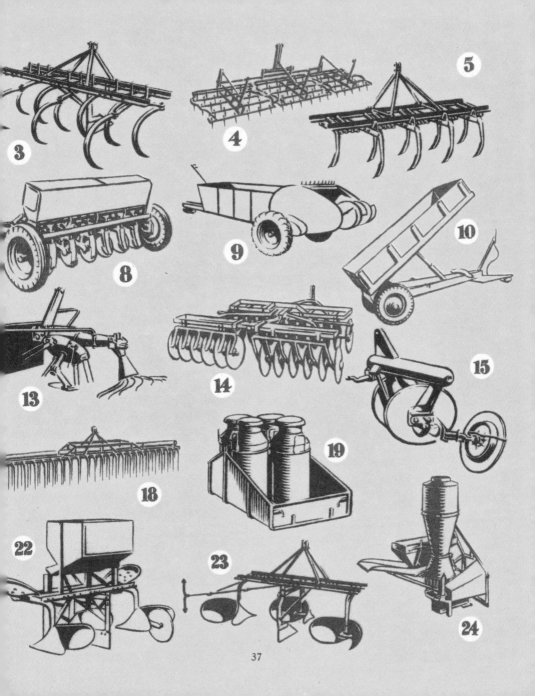

And as the sounds of the tractors echoed faintly from the far distance of the darkening fields, Peter and Pauline thought once more of all that they had heard about the Ferguson tractors and implements — how Mr. Ferguson had designed his tractor to fight the menace of world starvation, how each separate implement was skilfully integrated with the tractor to work as one complete unit, and how the tractor was manufactured at Coventry while the implements were made by different firms over the length and breadth of the country. They heard again their Uncle's voice telling them of the Ferguson Training School at Stoneleigh Abbey, and how Freddy Burbage had made the long trip all the way to Coventry to see something of the Ferguson Headquarters and Farming Exhibition. And finally, as four tired eyes were closing to the perfect sleep of healthy exhaustion, they thought of the work they'd done that day in the Wishing Well field . . . the Wishing Well field . . . the Wishing Well . . . Would they ever find the Wishing Well?

3.

FOR the next two weeks the twins were as busy as bees on the farm. They helped with this, they worked at that, they loaded this, they unloaded that — they took a Ferguson tractor here, they brought a Ferguson tractor there — they were tireless, excited and happy.

As each bright sun-filled day followed the other they grew as brown as berries and their hair was bleached to a dusky gold by the sun. The faithful Bess wouldn't leave them for a minute, but with tail wagging and joyous bark followed her heroes wherever they went.

But time, alas, slips on, until on one Friday morning the sad fact faced the twins that Mummy and Daddy would be coming to fetch them home again on the Sunday afternoon.

'Three weeks isn't long enough," grumbled Peter, moodily kicking his heels and watched by anxious Bess.

'Of course it isn't!' answered his sister — 'why we've hardly got used to the place!'

They were standing in a corner of the orchard, feeling very miserable and sad. The holiday was nearly over and in a few days time they would not only be home again but would be back at school. What a depressing thought on a nice Summer's day!

The Ferguson Training School at Stoneleigh Abbey.

THE TREASURE CHEST

'I know what we'll do,' said Peter suddenly, 'we'll take a tractor — there is one spare this morning — and go down to the Wishing Well field. We can play about in there all morning without damaging anything and it will be good fun too! Come on—I'll race you! First to the tractor drives first!'

So off they rushed with Bess lolloping behind them.

They arrived down at the field with Peter driving and Pauline clinging on to the back for dear life.

'Once round the field each," yelled Peter as he slowed down to let his sister jump off.

And then a strange thing happened. Hardly had he driven the Ferguson tractor two yards when — swish — the machine lurched to a swaying standstill. One of the rear wheels had slipped into a large hole.

Peter stopped the engine at once and climbed down to see what had happened.

The twins gasped in amazement.

They'd discovered the Wishing Well!

The mouth of the well had been boarded up at some time or other — perhaps fifty or even a hundred years ago — and as the years had passed by, the boards had slowly sunk down and become covered with grass-grown earth, until the well was entirely hidden in the green depths of the field. And now, by pure chance, Peter's tractor had betrayed the secret!

'Quick Peter!' shouted Pauline, 'drive the tractor off and let's have a look down the Well. There may be some water in it even yet. And at least we can have a wish!'

Peter did as he was told and the Ferguson climbed once more on to the level grass of the field. Behind it lay the deep dark hole of the newly discovered Wishing Well.

The twins — with Bess as well — peered cautiously over the rim of the Well and into the black depths below. Gradually — their eyes became accustomed to the darkness — they could perceive a dim yet exciting shape.

Pauline gasped.

'Why', she cried, 'it's a treasure-chest!' Sure enough deep at the bottom of the dried up waterless Well lay an old oak chest, studded with bolts and clasps and thick with the dust of the centuries.

Peter was off like a flash.

'I'm going to run and tell Uncle!' he yelled, and disappeared through the gate with a wave of his hand.

Pauline sat on the grass, and with Bess's wet nose in her lap she settled down to wait.

Nor had she to wait long.

Soon Uncle Arthur appeared at the wheel of his Ferguson tractor, with a trailer in which she could see Freddy Burbage, Peter and wonder of wonders! — old Gaffer Brownlow himself.

Uncle Arthur took the situation in at a glance as he peered down the well.

'H'mm,' said he, 'lucky that chest's got handles on each end. We'll shove a bar across the Well and lower a hook down on a cable. We can do it with a Ferguson and a winch. Quick, Freddy, get back to the farm for a winch!'

But it was half an hour before Freddy came back — he loved a gossip, he did — and when he *did* arrive he seemed to have brought Auntie Mabel and about half the village in his trailer. And who can blame them? You don't discover hidden treasure every day, do you?

Soon the cable was lowered, the chest handles securely hooked and the tractor started up. The chest was gradually drawn up to the mouth of the well, where willing hands dragged it on to the sun splashed grass.

It lay there, dirty, dusty and cobwebby, almost blinking at the fresh sweet daylight of the surface.

Then Uncle Arthur manfully applied his crowbar to the lock and with a creaking groan the lid wheezed open. The little band of treasure seekers gathered nearer — nobody wanted to miss anything and each was determined to see as much as he could.

There lying snugly inside the opened chest were beautifully worked silver candlesticks, great silver goblets, pewter tankards and magnificent enamelled trinkets — priceless treasures of a bygone day and age — glittering in God's daylight for the first time in almost four hundred years.

Peter and **Pauline** had discovered a treasure indeed, and the secret **whereabouts** of the mysterious Wishing Well were a secret no longer.

4.

ALL that long afternoon reporters and photographers from the big London newspapers had been driving through the white gates of Hollyhock Farm. They had taken pictures of Peter, Pauline, Uncle Arthur, Auntie Mabel, the Wishing Well, dog Bess (as excited as ever) and even old Gaffer Brownlow himself, who'd put on a carefully preserved celluloid collar for the occasion and pretended he was a 'hundred an' fower'!

The hamlet of Fernley Halt was agog with the news of the discovery. A large friendly police sergeant arrived on a bicycle and said he was sure that Uncle Arthur would be able to keep the treasure (since in England buried treasure is legally the property of the Crown and an inquest must be held over it). He also wanted to see the two nice children who had found it, and told them how clever they were.

Indeed, everyone made such a fuss and the house was filled with such comings and goings that Aunt Mabel said all the villagers could come to high tea on the lawn on the next day — which being Saturday was a very convenient arrangement.

. . . . .

So there, under the shade of the gracious old oak trees on the sun streaked lawn, the kindly village folk sat happily round the long trestle tables with the Master and Mistress of Hollyhock Farm.

In a place of honour at the table, enjoying themselves mightily, sat Peter and Pauline, with the devoted Bess in attendance as usual.

A really magnificent tea had been prepared with mounds of ham

sandwiches, bread and butter, tomatoes, trifles, custards, jellies, fruit cake and home made sausage rolls specially baked that morning. It was the last evening of the twins' stay at the farm and it was an evening they were not likely to forget.

That very morning their picture had been in all the papers. Mummy and Daddy had rung up to say how proud of them they were and Peter was now the possessor of a beautiful silver-handled knife while Pauline had chosen a silver and enamel box as her share of the treasure.

And so the evening wore on until the first faint stars began to glitter in purple heavens. One by one the guests rose to walk away along the sweet scented pathways through the fields which led to their homes, until at last only Uncle Arthur, Auntie Mabel and the twins were left.

'Time you were off to bed twins,' said Uncle Arthur, puffing contentedly at that luxurious last pipe of the evening.

'Yes, Uncle,' said Peter dutifully, 'but before we go, we would rather like to ask you something.'

'Of course,' nodded Uncle.

'Well, you see,' said Pauline, taking up the tale, 'we feel that it was our Ferguson which really found the treasure you know — and pulled it out of the well.'

Their Uncle smiled, and placed a kindly hand on each anxious golden head.

'My dear children,' he said, with a knowing look at Auntie Mabel just behind, 'that is the sole purpose of the Ferguson System throughout this great wide world of ours — to bring the true treasures up to us from out of the fruitful earth. For the earth is ever bountiful in its goodness, did men but realise it, and with such a System its riches may be more easily obtained than ever before in the history of mankind.'

And with those profound and well-chosen remarks, he led the little party back to the farmhouse, until at last only the tip of Bess's waving tail could be seen in the dusky star-lit distance.